Letts
gets you through

MATHS
IN 5 MINUTES A DAY

Chief editor: Zhou Jieying **Consultant:** Fan Lianghuo

CONTENTS

ZIFFIT

HOW TO USE THIS BOOK

The best way to help your child to build their confidence in maths and improve their number skills is to give them lots and lots of practice in the key facts and skills.

Written by maths experts, this series will help your child to become fluent in number facts, and help them to recall them quickly – both are essential for succeeding in maths.

This book provides ready-to-practise questions that comprehensively cover the number curriculum for Year 2. It contains 40 topic-based tests, each 5 minutes long, to help your child build up their mathematical fluency day-by-day.

Each test is divided into three Steps:

- **Step 1: Warm-up (1 minute)**
  This exercise helps your child to revise maths they should already know and gives them preparation for Step 2.
- **Step 2: Rapid calculation ($2\frac{1}{2}$ minutes)**
  This exercise is a set of questions focused on the topic area being tested.
- **Step 3: Challenge ($1\frac{1}{2}$ minutes)**
  This is a more testing exercise designed to stretch your child's mental abilities.

Some of the tests also include a Tip to help your child answer questions of a particular type.

Your child should attempt to answer as many questions as possible in the time allowed at each Step. Answers are provided at the back of the book.

To help to measure progress, each test includes boxes for recording the date of the test, the total score obtained and the total time taken.

ACKNOWLEDGEMENTS

The authors and publisher are grateful to the copyright holders for permission to use quoted materials and images.

All images are © HarperCollins*Publishers* Ltd and © Shutterstock.com

Every effort has been made to trace copyright holders and obtain their permission for the use of copyright material. The authors and publisher will gladly receive information enabling them to rectify any error or omission in subsequent editions. All facts are correct at time of going to press.

Published by Letts Educational in association with East China Normal University Press

Letts Educational
An imprint of HarperCollins*Publishers*
1 London Bridge Street
London SE1 9GF

ISBN: 978-0-00-831109-4

First published 2019

10 9 8 7 6 5 4 3 2 1

British Library Cataloguing in Publication Data.

A CIP record of this book is available from the British Library.

Publisher: Fiona McGlade
Consultant: Fan Lianghuo
Authors: Zhou Jieying, Xu Jing and Yao Lu
Editors: Ni Ming and Xu Huiping
Contributor: Paul Hodge
Project Management and Editorial: Richard Toms, Lauren Murray and Marie Taylor
Cover Design: Sarah Duxbury
Inside Concept Design: Paul Oates and Ian Wrigley
Layout: Jouve India Private Limited

Printed in Great Britain by Martins the Printers

MIX
Paper from responsible source
FSC™ C007454

This book is produced from independently certified FSC™ paper to ensure responsible forest management.

For more information visit:
www.harpercollins.co.uk/green

1 Addition and subtraction of numbers up to 20

Date: 8.4.20
Day of Week: Wednesday

STEP 1 (1 min) Warm-up

Start the timer

Answer these.

8 + 5 = 13 ✓	11 − 2 = 9 ✓	6 + 9 = 15 ✓	10 − 8 = 2 ✓
12 − 6 = 6 ✓	0 + 13 = 13 ✓	15 + 5 = 20 ✓	20 − 10 = 10 ✓
9 + 0 + 8 = 17 ✓	15 − 8 − 2 = 5 ✓		

STEP 2 (2.5 min) Rapid calculation

Answer these.

14 + 6 = ☐	17 − 8 = ☐	9 + 8 = ☐	18 + 0 = ☐
7 + 5 = ☐	4 + 17 = ☐	9 + 11 = ☐	16 − 8 = ☐
8 + 10 = ☐	16 − 4 = ☐	11 + 0 = ☐	15 − 15 = ☐
15 − 0 = ☐	11 + 8 = ☐	15 − 8 = ☐	7 + 7 = ☐
17 − 9 = ☐	2 + 5 + 8 = ☐	7 + 5 − 7 = ☐	8 − 6 + 8 = ☐

STEP 3 (1.5 min) Challenge

Fill in the missing numbers.

8 + ☐ = 16	☐ − 6 = 5	4 + ☐ = 13	14 − ☐ = 6
☐ + 10 = 12	7 + ☐ = 13	☐ − 7 = 7	15 − ☐ = 9

Time spent: ______ min ______ sec. Total: ______ out of 38

Date: ______________________

Day of Week: ______________________

STEP 1 (1 min) Warm-up

Start the timer

Answer these.

$5 + 10 = \square$ $13 - 3 = \square$ $10 + 9 = \square$ $17 - 7 = \square$

$7 + 7 = \square$ $10 + 5 = \square$ $10 + 3 = \square$ $19 - 9 = \square$

$17 - 10 = \square$ $14 - 7 = \square$

STEP 2 (2.5 min) Rapid calculation

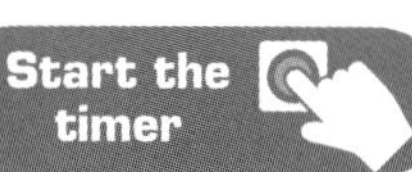

Fill in each box with **>**, **<** or **=**.

$14 - 8 \ \square \ 14$ $3 + 8 \ \square \ 8$ $5 + 5 \ \square \ 10$ $11 - 6 \ \square \ 11$

$11 \ \square \ 10 + 3$ $11 \ \square \ 10 + 2$ $11 \ \square \ 10 + 1$ $11 \ \square \ 10 + 0$

$11 \ \square \ 12 - 0$ $11 \ \square \ 12 - 1$ $11 \ \square \ 12 - 2$ $11 \ \square \ 12 - 3$

$11 \ \square \ 11 + 0$ $11 \ \square \ 11 - 0$ $11 \ \square \ 11 - 1$ $11 \ \square \ 11 + 1$

STEP 3 (1.5 min) Challenge

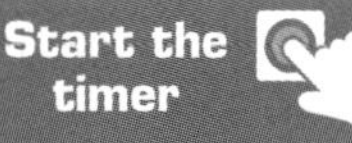

Find the largest number or digit that can be filled in each box.

$\square 2 < 16$ $9 \square < 100$ $\square 7 < 98$ $65 > 6 \square$

$\square + 2 < 16$ $9 + \square < 15$ $\square - 3 < 13$ $17 - \square > 6$

Time spent: ________ min ________ sec. Total: ________ out of 34

3 Adding a whole tens to a two-digit number (1)

Date: ______________________

Day of Week: ______________________

STEP 1 (1 min) Warm-up

Start the timer

Answer these.

30 + 40 = ☐	70 + 20 = ☐	50 + 30 = ☐	80 + 10 = ☐
70 + 2 = ☐	90 + 5 = ☐	80 + 4 = ☐	90 + 7 = ☐

STEP 2 (2.5 min) Rapid calculation

Answer these.

32 + 40 = ☐	75 + 20 = ☐	54 + 30 = ☐	87 + 10 = ☐
77 + 20 = ☐	58 + 30 = ☐	30 + 63 = ☐	30 + 47 = ☐
25 + 60 = ☐	51 + 40 = ☐	10 + 39 = ☐	52 + 30 = ☐
60 + 12 = ☐	40 + 15 = ☐	39 + 10 = ☐	30 + 25 = ☐
70 + 13 = ☐	30 + 56 = ☐	51 + 20 = ☐	50 + 50 = ☐

STEP 3 (1.5 min) Challenge

Fill in the missing numbers.

30 + ☐ = 60	☐ + 40 = 80	☐ + 20 = 40
30 + ☐ = 78	7 + ☐ = 67	☐ + 40 = 69
25 + ☐ = 55	30 + ☐ = 65	98 = ☐ + 50

Time spent: ________ min ________ sec. Total: ________ out of 37

Date: ______________________

Day of Week: ________________

Adding a whole tens to a two-digit number (2)

STEP 1 (1 min) Warm-up

Start the timer

Add these amounts of money.

£10 + £7 =	16p + 20p =
£40 + £6 =	30p + 18p =
£3 + £80 =	22p + 70p =
£9 + £50 =	10p + 87p =
£60 + £5 =	40p + 30p =

STEP 2 (2.5 min) Rapid calculation

1. Add these lengths.

17 m + 40 m =	50 cm + 17 cm =
15 m + 60 m =	44 cm + 40 cm =
10 m + 75 m =	12 cm + 70 cm =
43 m + 20 m =	35 cm + 30 cm =

2. Fill in each box with **>**, **<** or **=**.

£26 + £50 ☐ £77　　£43 + £20 ☐ £63　　£38 + £50 ☐ £89

£72 + £20 ☐ £81　　21p + 20p ☐ 41p　　54p + 40p ☐ 93p

57p + 30p ☐ 87p　　17p + 80p ☐ 98p

STEP 3 (1.5 min) Challenge

Fill in the missing lengths.

36 m + ☐ = 56 m　　☐ + 20 m = 62 m　　44 m + ☐ = 84 m

61 cm + ☐ = 91 cm　　50 cm + ☐ = 99 cm　　☐ + 27 cm = 67 cm

55 m + ☐ = 65 m　　22 cm + ☐ = 42 cm　　86 m = ☐ + 30 m

Time spent: ________ min ________ sec. Total: ________ out of 35

5 Subtracting a whole tens from a two-digit number (1)

Date: ____________________

Day of Week: ____________________

STEP 1 (1 min) Warm-up

Start the timer

Answer these.

60 – 20 = □	80 – 10 = □	90 – 70 = □	50 – 20 = □
40 + 1 = □	70 + 7 = □	20 + 6 = □	30 + 4 = □

STEP 2 (2.5 min) Rapid calculation

Answer these.

61 – 20 = □	87 – 10 = □	96 – 70 = □	54 – 20 = □
77 – 20 = □	58 – 30 = □	63 – 30 = □	47 – 20 = □
65 – 20 = □	51 – 40 = □	39 – 10 = □	52 – 30 = □
62 – 60 = □	95 – 50 = □	85 – 40 = □	35 – 30 = □
97 – 90 = □	91 – 80 = □	51 – 20 = □	50 – 50 = □

STEP 3 (1.5 min) Challenge

Fill in the missing numbers.

93 – □ = 60	□ – 50 = 28	□ – 70 = 12
86 – □ = 16	45 – □ = 5	□ – 50 = 46
88 – □ = 38	99 – □ = 69	37 = □ – 20

Time spent: ______ min ______ sec. Total: ______ out of 37

Date: ______________________

Day of Week: ______________________

Subtracting a whole tens from a two-digit number (2)

STEP 1 (1 min) Warm-up

Start the timer

Subtract these amounts.

£30 – £5 =	77p – 20p =
£40 – £4 =	83p – 40p =
£60 – £2 =	48p – 20p =
£70 – £6 =	95p – 80p =
£50 – £9 =	64p – 50p =

STEP 2 (2.5 min) Rapid calculation

1. Subtract these lengths.

42 m – 30 m =	57 cm – 30 cm =
88 m – 50 m =	96 cm – 70 cm =
69 m – 10 m =	35 cm – 30 cm =
71 m – 60 m =	99 cm – 80 cm =

2. Fill in each box with **>**, **<** or **=**.

£47 – £10 ☐ £38   £36 – £20 ☐ £15   £71 – £40 ☐ £31

£95 – £70 ☐ £35   78p – 30p ☐ 49p   65p – 50p ☐ 5p

64p – 40p ☐ 24p   88p – 80p ☐ 9p

STEP 3 (1.5 min) Challenge

Fill in the missing lengths.

34 m – ☐ = 8 m   55 m – ☐ = 28 m   ☐ – 40 m = 29 m

86 cm – ☐ = 33 cm   ☐ – 40 cm = 22 cm   ☐ – 60 cm = 33 cm

67 m – ☐ = 24 m   66 cm – ☐ = 48 cm   35 m = ☐ – 20 m

Time spent: ________ min ________ sec. Total: ________ out of 35

7 Adding a one-digit number to a two-digit number

Date: ______________________

Day of Week: ______________________

STEP 1 (1 min) Warm-up

Start the timer

Answer these.

6 + 2 = ☐ 5 + 4 = ☐ 2 + 5 = ☐ 3 + 6 = ☐

40 + 20 = ☐ 70 + 10 = ☐ 30 + 50 = ☐ 7 + 40 = ☐

STEP 2 (2.5 min) Rapid calculation

Start the timer

Answer these.

83 + 6 = ☐ 51 + 4 = ☐ 70 + 9 = ☐ 63 + 6 = ☐

6 + 52 = ☐ 41 + 8 = ☐ 7 + 82 = ☐ 6 + 33 = ☐

5 + 33 = ☐ 3 + 66 = ☐ 51 + 6 = ☐ 41 + 5 = ☐

60 + 8 = ☐ 20 + 9 = ☐ 40 + 7 = ☐ 50 + 9 = ☐

66 + 2 = ☐ 25 + 4 = ☐ 42 + 5 = ☐ 53 + 6 = ☐

STEP 3 (1.5 min) Challenge

Fill in the missing numbers.

5 + ☐ = 65 ☐ + 8 = 98 ☐ + 9 = 69

6 + ☐ = 77 7 + ☐ = 58 ☐ + 4 = 47

5 + ☐ = 45 3 + ☐ = 84 58 = ☐ + 52

Time spent: ________ min ________ sec. Total: ________ out of 37

Date: ______________________

Day of Week: ______________________

Subtracting a one-digit number from a two-digit number

STEP 1 (1 min) Warm-up

Start the timer

Answer these.

7 – 4 = ☐	9 – 8 = ☐	8 – 3 = ☐	6 – 4 = ☐
50 + 3 = ☐	70 + 1 = ☐	40 + 5 = ☐	90 + 2 = ☐

STEP 2 (2.5 min) Rapid calculation

Start the timer

Answer these.

20 – 6 = ☐	30 – 7 = ☐	48 – 9 = ☐	53 – 8 = ☐
56 – 5 = ☐	82 – 5 = ☐	64 – 6 = ☐	73 – 4 = ☐
61 – 2 = ☐	62 – 3 = ☐	63 – 4 = ☐	64 – 5 = ☐
61 – 9 = ☐	62 – 8 = ☐	63 – 7 = ☐	64 – 6 = ☐
57 – 4 = ☐	79 – 8 = ☐	48 – 3 = ☐	96 – 4 = ☐

STEP 3 (1.5 min) Challenge

Fill in the missing numbers.

51 – ☐ = 9	82 – ☐ = 5	46 – ☐ = 8
62 – ☐ = 7	64 – ☐ = 8	98 – ☐ = 8
3 = 91 – ☐	5 = 73 – ☐	72 – ☐ = 62 + 8

Time spent: ______ min ______ sec. Total: ______ out of 37

9 Adding or subtracting a one-digit number to or from a two-digit number

Date: ____________________

Day of Week: ____________________

STEP 1 (1 min) Warm-up

Start the timer

Answer these.

$53 + 8 = \square$ $31 + 6 = \square$ $46 + 5 = \square$ $76 + 8 = \square$

$92 + 0 = \square$ $30 - 2 = \square$ $56 - 7 = \square$ $75 - 9 = \square$

STEP 2 (2.5 min) Rapid calculation

Answer these.

$67 - 8 = \square$ $63 + 5 = \square$ $28 + 7 = \square$ $73 - 5 = \square$

$43 - 9 = \square$ $25 + 9 = \square$ $66 + 7 = \square$ $90 - 9 = \square$

$70 - 0 = \square$ $70 + 0 = \square$ $78 - 8 = \square$ $51 + 9 = \square$

$73 - 7 = \square$ $50 - 2 = \square$ $27 + 3 = \square$ $64 + 6 = \square$

$81 - 4 = \square$ $35 - 0 = \square$

STEP 3 (1.5 min) Challenge

Fill in the missing numbers.

$35 + \square = 40$ $\square - 35 = 5$ $40 - \square = 35$

$\square - 8 = 65$ $\square + 76 = 85$ $72 - \square = 65$

$53 = \square + 47$ $17 = 24 - \square$ $\square - 22 = 60$

Time spent: ________ min ________ sec. Total: ________ out of 35

Date: ______________________

Day of Week: ______________________

Adding a two-digit number to a two-digit number

10

STEP 1 (1 min) Warm-up

Start the timer

Answer these.

3 + 4 = ☐ 5 + 3 = ☐ 30 + 50 = ☐ 20 + 60 = ☐

40 + 50 = ☐ 20 + 70 = ☐ 10 + 30 = ☐ 34 + 52 = ☐

STEP 2 (2.5 min) Rapid calculation

Start the timer

Answer these.

21 + 70 = ☐ 30 + 57 = ☐ 50 + 35 = ☐ 74 + 10 = ☐

41 + 54 = ☐ 12 + 73 = ☐ 21 + 24 = ☐ 64 + 15 = ☐

42 + 25 = ☐ 23 + 73 = ☐ 86 + 11 = ☐ 71 + 13 = ☐

21 + 21 = ☐ 43 + 43 = ☐ 34 + 34 = ☐ 13 + 13 = ☐

21 + 68 = ☐ 43 + 54 = ☐ 25 + 73 = ☐ 14 + 32 = ☐

STEP 3 (1.5 min) Challenge

Start the timer

Fill in the missing numbers.

11 + 22 = ☐ 22 + 33 = ☐ 33 + 44 = ☐

32 + 23 = ☐ 61 + 16 = ☐ 27 + 72 = ☐

18 + ☐ = 88 51 + ☐ = 79 66 = ☐ + 44

Time spent: ______ min ______ sec. Total: ______ out of 37

11 Subtracting a two-digit number from a two-digit number

Date: ____________

Day of Week: ____________

STEP 1  1 min Warm-up

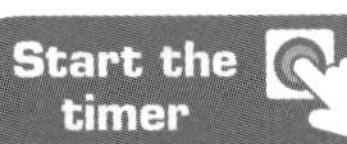

Answer these.

80 – 70 = □ 90 – 30 = □ 70 – 20 = □ 60 – 40 = □

50 – 30 = □ 86 – 74 = □ 99 – 35 = □ 76 – 23 = □

STEP 2 2.5 min Rapid calculation

Answer these.

56 – 10 = □ 38 – 18 = □ 50 – 10 = □ 97 – 17 = □

54 – 41 = □ 73 – 12 = □ 66 – 33 = □ 84 – 42 = □

96 – 25 = □ 68 – 23 = □ 86 – 21 = □ 78 – 55 = □

79 – 26 = □ 43 – 43 = □ 49 – 10 = □ 62 – 21 = □

67 – 42 = □ 58 – 37 = □

STEP 3 1.5 min Challenge

Fill in the missing numbers.

88 – 22 = □ 77 – 33 = □ 66 – 44 = □

88 – 15 = □ 78 – 25 = □ 68 – 35 = □

53 – □ = 3 □ – 18 = 70 57 = 98 – □

Time spent: ______ min ______ sec. Total: ______ out of 35

Addition and subtraction of two-digit numbers (1)

STEP 1  Warm-up

Start the timer

Answer these.

53 + 20 = ☐	27 – 8 = ☐	40 – 14 = ☐	67 + 30 = ☐
40 + 55 = ☐	81 – 60 = ☐	96 – 9 = ☐	89 + 5 = ☐

STEP 2 2.5 min Rapid calculation

Answer these.

21 + 78 = ☐	57 – 27 = ☐	53 + 35 = ☐	74 – 24 = ☐
36 + 54 = ☐	17 + 73 = ☐	50 – 29 = ☐	63 – 36 = ☐
56 + 25 = ☐	73 – 18 = ☐	65 + 25 = ☐	78 – 13 = ☐
51 – 51 = ☐	49 + 18 = ☐	75 – 57 = ☐	68 + 10 = ☐
90 – 18 = ☐	40 + 40 = ☐		

STEP 3 1.5 min Challenge

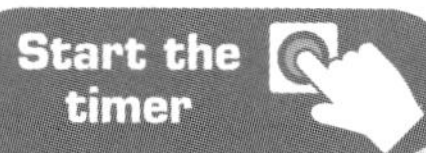

Fill in the missing numbers.

67 – 18 = ☐	38 + 13 = ☐	16 + ☐ = 45
18 + 45 = ☐	85 – 59 = ☐	71 – ☐ = 35
93 – 78 = ☐	27 + 25 = ☐	35 = ☐ – 35

Time spent: ________ min ________ sec. Total: ________ out of 35

13 Addition and subtraction of two-digit numbers (2)

Date: ____________________

Day of Week: ____________________

STEP 1 (1 min) Warm-up

Answer these.

£34 + £16 =	38p – 19p =
£18 + £22 =	65p – 38p =
£45 + £25 =	55p – 26p =
£17 + £53 =	72p – 33p =
£51 + £39 =	93p – 88p =

STEP 2 (2.5 min) Rapid calculation

Answer these.

24 m + 34 m =	56 m – 36 m =
75 m – 37 m =	63 m + 30 m =
84 cm – 36 cm =	50 cm + 25 cm =
33 cm – 18 cm =	48 cm + 27 cm =
54 m + 36 m =	45 m – 26 m =
69 m + 15 m =	35 m + 35 m =
74 cm + 17 cm =	38 cm + 40 cm =
86 cm – 29 cm =	98 cm – 77 cm =
46 cm + 19 cm =	73 cm – 18 cm =

STEP 3 (1.5 min) Challenge

Fill in the missing amounts.

☐ – £34 = £26   £45 + ☐ = £90   £66 – ☐ = £33

☐ – £54 = £18   £37 – ☐ = £19   ☐ + £44 = £82

97p = ☐ + 14p + 22p   55p = ☐ – 18p – 14p   33p + ☐ = 72p – 15p

Time spent: ________ min ________ sec. Total: ________ out of 37

Date: ______________________

Day of Week: ______________________

Adding three numbers

STEP 1 (1 min) Warm-up

Start the timer

Answer these.

3 + 5 + 5 = ☐ 8 + 2 + 3 = ☐ 1 + 4 + 6 = ☐ 7 + 2 + 3 = ☐

3 + 9 + 1 = ☐ 5 + 5 + 6 = ☐ 8 + 5 + 2 = ☐ 5 + 4 + 5 = ☐

STEP 2 (2.5 min) Rapid calculation

Answer these.

9 + 1 + 1 = ☐ 8 + 2 + 2 = ☐ 4 + 3 + 5 = ☐ 3 + 1 + 7 = ☐

7 + 4 + 8 = ☐ 7 + 5 + 6 = ☐ 7 + 3 + 4 = ☐ 1 + 9 + 5 = ☐

5 + 7 + 5 = ☐ 4 + 8 + 6 = ☐ 4 + 8 + 7 = ☐ 3 + 8 + 6 = ☐

7 + 5 + 3 = ☐ 3 + 7 + 8 = ☐ 7 + 7 + 1 = ☐ 6 + 6 + 4 = ☐

5 + 4 + 4 = ☐ 3 + 3 + 6 = ☐

STEP 3 (1.5 min) Challenge

Answer these.

9 + 7 + 9 = ☐ 8 + 8 + 9 = ☐

9 + 7 + 8 = ☐ 7 + 7 + 6 = ☐

8 + 6 + 8 = ☐ 9 + 9 + 9 = ☐

7 + 7 + 7 = ☐ 8 + 8 + 8 = ☐

Time spent: ________ min ________ sec. Total: ________ out of 34

15 General review (1)

Date: ____________________

Day of Week: ____________________

STEP 1 (1 min) Warm-up

Start the timer

Answer these.

21 + 14 = □   63 + 25 = □   46 + 33 = □   64 + 23 = □

40 – 35 = □   85 – 23 = □   78 – 44 = □   21 + 19 = □

STEP 2 (2.5 min) Rapid calculation

Answer these.

44 + 28 = □   78 + 18 = □   62 – 35 = □   71 – 52 = □

72 – 28 = □   96 – 18 = □   27 + 35 = □   19 + 52 = □

54 – 28 = □   81 – 67 = □   61 – 44 = □   57 – 29 = □

83 – 79 = □   38 + 43 = □   42 – 19 = □   29 + 37 = □

63 + 28 = □   42 – 37 = □

STEP 3 (1.5 min) Challenge

Fill in the missing numbers.

□ + 18 = 40   29 + □ = 51   62 = 36 + □

□ + 24 = 42   72 + □ = 100   29 = 82 – □

74 – □ = 58   □ – 36 = 27   48 = □ – 27

Time spent: ________ min ________ sec. Total: ________ out of 35

Date: ________________________

Day of Week: ____________________

STEP 1 (1 min) Warm-up

Start the timer

Answer these.

24 + 38 = ☐	55 – 29 = ☐	46 + 39 = ☐	56 – 48 = ☐
80 – 46 = ☐	91 – 58 = ☐	26 + 39 = ☐	65 – 26 = ☐

STEP 2 (2.5 min) Rapid calculation

Start the timer

Fill in the missing numbers.

16 + ☐ = 47	☐ + 26 = 75	☐ + 27 = 55	☐ + 37 = 76
14 + ☐ = 68	56 + ☐ = 71	56 + ☐ = 83	☐ + 25 = 72
15 + ☐ = 53	☐ + 23 = 48	21 + ☐ = 50	☐ + 36 = 84
25 + ☐ = 91	38 + ☐ = 97	☐ + 52 = 86	☐ + 44 = 80

STEP 3 (1.5 min) Challenge

Start the timer

Fill in each box with **>**, **<** or **=**.

8 + 7 + 5 ☐ 34 – 13	3 + 9 + 6 ☐ 44 – 27
4 + 8 + 9 ☐ 9 + 12	6 + 9 + 9 ☐ 43 – 20
8 + 6 + 7 ☐ 84 – 62	8 + 8 + 7 ☐ 92 – 69
8 + 9 + 9 ☐ 9 + 16	9 + 9 + 9 ☐ 15 + 13

Time spent: ________ min ________ sec. Total: ________ out of 32

17 General review (3)

Date: ____________

Day of Week: ____________

STEP 1 (1 min) Warm-up

Start the timer

Answer these.

£5 + £7 + £9 = ☐ | £7 + £6 + £5 = ☐ | £3 + £5 + £7 = ☐

£2 + £8 + £8 = ☐ | 4p + 4p + 5p = ☐ | 6p + 7p + 7p = ☐

9p + 4p + 5p = ☐ | 8p + 5p + 4p = ☐

STEP 2 (2.5 min) Rapid calculation

Fill in each box with >, < or =.

10 m + 6 m + 6 m ☐ 41 m – 20 m | 7 m + 10 m + 7 m ☐ 53 m – 28 m

8 m + 5 m + 10 m ☐ 67 m – 45 m | 5 cm + 10 cm + 9 cm ☐ 62 cm – 37 cm

10 cm + 8 cm + 8 cm ☐ 91 cm – 65 cm | 9 cm + 9 cm + 10 cm ☐ 85 cm – 58 cm

9 m + 5 m + 9 m ☐ 81 m – 58 m | 6 cm + 8 cm + 3 cm ☐ 99 cm – 81 cm

STEP 3 (1.5 min) Challenge

Fill in the missing amounts.

£95 – ☐ = £32 + £16 | ☐ – £35 = £54 – £21

☐ – £31 = £19 + £45 | £71 – ☐ = £65 – £23

66p – ☐ = 12p + 35p | ☐ – 27p = 58p – 14p

83p – ☐ = 92p – 61p | ☐ – 37p = 18p + 28p

Time spent: ______ min ______ sec. Total: ______ out of 24

Date: ______________________

Day of Week: ______________________

General review (4) 18

STEP 1 (1 min) Warm-up

Start the timer

A car journey is completed in two stages. What is the total distance travelled in each journey?

26 km and 45 km Total: ☐ km

34 km and 38 km Total: ☐ km

45 km and 29 km Total: ☐ km

53 km and 19 km Total: ☐ km

65 km and 28 km Total: ☐ km

47 km and 47 km Total: ☐ km

STEP 2 (2.5 min) Rapid calculation

The money in three piggy banks is added together. What is the total amount in each piggy bank?

£5, £7 and £8 Total: £ ☐

£8, £6 and £4 Total: £ ☐

£3, £7 and £10 Total: £ ☐

£9, £5 and £6 Total: £ ☐

£10, £8 and £8 Total: £ ☐

£9, £10 and £9 Total: £ ☐

£6, £9 and £9 Total: £ ☐

£4, £8 and £7 Total: £ ☐

£6, £6 and £5 Total: £ ☐

£3, £10 and £9 Total: £ ☐

STEP 3 (1.5 min) Challenge

Fill in the missing numbers.

1. Harry has £37. He wants to save £74. How much more money does he need? £ ☐

2. Leona has £44. She wants to save £92. How much more money does she need? £ ☐

3. Faisal has £59. He wants to save £97. How much more money does he need? £ ☐

4. Lara has £18. She wants to save £85. How much more money does she need? £ ☐

Time spent: ______ min ______ sec. Total: ______ out of 20

19 Commutative law of addition

Date: ____________________

Day of Week: ____________________

STEP 1 (1 min) Warm-up

Start the timer

Answer these.

37 + 48 = ☐   48 + 37 = ☐   68 + 27 = ☐   27 + 68 = ☐

73 + 19 = ☐   19 + 73 = ☐   45 + 47 = ☐   47 + 45 = ☐

39 + 55 = ☐   55 + 39 = ☐   49 + 37 = ☐   37 + 49 = ☐

STEP 2 (2.5 min) Rapid calculation

Fill in the missing numbers.

1. 25 + 34 = 34 + ☐ = ☐
2. 67 + 16 = ☐ + 67 = ☐
3. 58 + 37 = 37 + ☐ = ☐
4. 47 + 25 = 25 + ☐ = ☐
5. 59 + 24 = ☐ + 59 = ☐
6. 29 + 47 = 47 + ☐ = ☐
7. 23 + 68 = ☐ + ☐ = ☐
8. 75 + 18 = ☐ + ☐ = ☐
9. 43 + 35 = ☐ + ☐ = ☐
10. 18 + 68 = ☐ + ☐ = ☐
11. 83 + 16 = ☐ + ☐ = ☐
12. 37 + 44 = ☐ + ☐ = ☐

STEP 3 (1.5 min) Challenge

Fill in the missing numbers row by row.

1. 57 + 26 = ☐   56 + 27 = ☐   54 + 29 = ☐   52 + 31 = ☐
2. 37 + 49 = ☐   39 + 47 = ☐   33 + 53 = ☐   ☐ + 59 = ☐
3. 76 + 22 = ☐   66 + 32 = ☐   52 + 46 = ☐   40 + ☐ = ☐
4. 52 + 39 = ☐   ☐ + 49 = ☐   ☐ + ☐ = ☐   ☐ + ☐ = ☐

Time spent: ________ min ________ sec. Total: ________ out of 40

Date: ______________________

Day of Week: ______________________

Odd and even numbers 20

STEP 1 (1 min) Warm-up

Start the timer

1. Circle the odd numbers.

13	44	21	12	36	46	8	35	47	54
49	50	38	34	22	85	79	57	68	90

2. Circle the even numbers.

88	37	46	21	89	54	42	92	98	65
34	53	64	74	67	70	31	2	23	100

STEP 2 (2.5 min) Rapid calculation

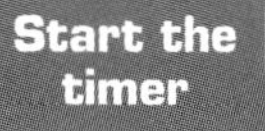

1. Write all the odd numbers between:

4 and 10 ______________ 14 and 20 ______________

32 and 38 ______________ 50 and 60 ______________

75 and 85 ______________ 83 and 93 ______________

2. Write all the even numbers between:

1 and 7 ______________ 13 and 19 ______________

24 and 30 ______________ 40 and 50 ______________

64 and 74 ______________ 88 and 98 ______________

STEP 3 (1.5 min) Challenge

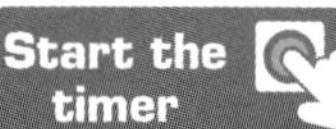

1. 45 78 24 82 47 64 27 44 20 76 65 51

Greatest even number = ☐ Smallest odd number = ☐

2. 34 53 10 87 68 75 17 33 59 73 24 35

Greatest even number = ☐ Smallest odd number = ☐

3. 23 15 67 34 71 59 75 64 31 87 57 85

Greatest even number = ☐ Smallest odd number = ☐

4. 61 77 43 87 49 71 80 73 40 74 97 48

Greatest even number = ☐ Smallest odd number = ☐

Time spent: ________ min ________ sec. Total: ________ out of 40

21 Introduction to multiplication

Date: ____________________

Day of Week: ____________________

STEP 1 (1 min) Warm-up

Answer these.

2 + 2 + 2 = ☐ 3 + 3 + 3 = ☐ 4 + 4 + 4 + 4 = ☐ 5 + 5 + 5 + 5 = ☐

6 + 6 + 6 = ☐ 7 + 7 + 7 = ☐ 3 + 3 + 3 + 3 = ☐ 6 + 6 + 6 + 6 = ☐

4 + 4 + 4 = ☐ 9 + 9 + 9 = ☐ 7 + 7 + 7 + 7 = ☐ 8 + 8 + 8 + 8 + 8 = ☐

STEP 2 (2.5 min) Rapid calculation

Fill in the missing numbers.

2 + 2 + 2 + 2 = **4** × 2 = ☐ 4 + 4 + 4 = ☐ × 4 = ☐

3 + 3 + 3 + 3 = ☐ × 3 = ☐ 1 + 1 + 1 + 1 + 1 = ☐ × 1 = ☐

5 + 5 + 5 + 5 = ☐ × 5 = ☐ 7 + 7 + 7 = ☐ × 7 = ☐

9 + 9 + 9 = ☐ × 9 = ☐ 8 + 8 + 8 + 8 = ☐ × 8 = ☐

6 + 6 + 6 + 6 + 6 = ☐ × ☐ = ☐ 7 + 7 + 7 + 7 + 7 = ☐ × ☐ = ☐

1 + 1 + 1 + 1 + 1 + 1 + 1 = ☐ × ☐ = ☐ 3 + 3 + 3 = ☐ × ☐ = ☐

5 + 5 + 5 + 5 + 5 + 5 + 5 + 5 = ☐ × ☐ = ☐ 8 + 8 = ☐ × ☐ = ☐

STEP 3 (1.5 min) Challenge

Fill in the missing numbers.

3 + 3 + 3 + 3 + 3 + 3 + 3 + 3 + 3 = ☐ × ☐ = ☐

8 + 8 + 8 + 8 + 8 + 8 + 8 = ☐ × ☐ = ☐

10 + 10 + 10 + 10 = ☐ × ☐ = ☐

9 + 9 + 9 + 9 + 9 + 9 + 9 + 9 = ☐ × ☐ = ☐

12 + 12 + 12 + 12 + 12 + 12 + 12 + 12 = ☐ × ☐ = ☐

Time spent: ________ min ________ sec. Total: ________ out of 31

Date:

Day of Week:

Times 22

STEP 1 (1 min) Warm-up

Start the timer

Answer these.

2 + 2 + 2 = ☐

10 + 10 + 10 + 10 = ☐

10 + 10 + 10 = ☐

10 + 10 + 10 + 10 + 10 = ☐

5 + 5 + 5 + 5 + 5 + 5 = ☐

5 + 5 + 5 + 5 = ☐

5 + 5 + 5 = ☐

2 + 2 + 2 + 2 = ☐

2 + 2 + 2 + 2 + 2 + 2 = ☐

10 + 10 + 10 + 10 + 10 + 10 + 10 = ☐

STEP 2 (2.5 min) Rapid calculation

Complete the table. The first row has been done for you.

5 + 5 = 10	**two 5s**	**2 times 5**	**2 × 5 = 10**
2 + 2 + 2 + 2 + 2 = 10			
10 + 10 + 10 + 10 = 40			
2 + 2 + 2 + 2 + 2 + 2 + 2 + 2 = 16			
10 + 10 + 10 + 10 + 10 + 10 = 60			
5 + 5 + 5 + 5 + 5 = 25			

STEP 3 (1.5 min) Challenge

Complete the table. The first row has been done for you.

2 + 2 + 2 + 2 = **8**	**4 × 2 = 8**	**2 × 4 = 8**
5 + 5 =		
10 + 10 + 10 =		
2 + 2 + 2 + 2 + 2 =		
5 + 5 + 5 + 5 + 5 + 5 + 5 =		
10 + 10 + 10 + 10 + 10 + 10 + 10 =		
5 + 5 + 5 + 5 + 5 + 5 + 5 + 5 =		
2 + 2 + 2 + 2 + 2 + 2 + 2 + 2 + 2 + 2 =		

Time spent: ______ min ______ sec. Total: ______ out of 46

23 Multiplication as an array

Date: ______________

Day of Week: ______________

STEP 1  Warm-up

Start the timer

Complete the table. The first row has been done for you.

Array	Number of rows	Number of columns	Total number of dots
	2	3	6

STEP 2 2.5 min Rapid calculation

Complete two multiplications for each array. The first one has been done for you.

2 × 4 = 8					
4 × 2 = 8					

STEP 3 1.5 min Challenge

How many coins are there in:

4 rows of 2 coins? ☐ 6 rows of 5 coins? ☐ 7 rows of 10 coins? ☐

2 rows of 9 coins? ☐ 8 rows of 5 coins? ☐ 9 rows of 10 coins? ☐

Time spent: ______ min ______ sec. Total: ______ out of 31

Date: ____________

Day of Week: ____________

Multiplication of 10

STEP 1 (1 min) Warm-up

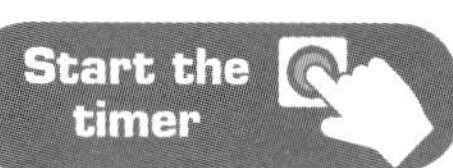

Answer these.

10 + 10 = ☐ 80 – 10 = ☐ 60 – 10 = ☐ 90 – 10 = ☐

10 + 40 = ☐ 50 + 10 = ☐ 10 + 60 = ☐ 10 + 70 = ☐

10 + 10 + 10 = ☐ 10 + 0 + 10 = ☐ 10 + 10 + 10 + 10 = ☐

STEP 2 (2.5 min) Rapid calculation

Answer these.

1 × 10 = ☐ 2 × 10 = ☐ 3 × 10 = ☐ 4 × 10 = ☐

5 × 10 = ☐ 6 × 10 = ☐ 7 × 10 = ☐ 8 × 10 = ☐

9 × 10 = ☐ 10 × 10 = ☐ 10 × 1 = ☐ 10 × 2 = ☐

10 × 3 = ☐ 10 × 4 = ☐ 10 × 5 = ☐ 10 × 6 = ☐

10 × 11 = ☐ 10 × 8 = ☐ 10 × 12 = ☐ 10 × 0 = ☐

STEP 3 (1.5 min) Challenge

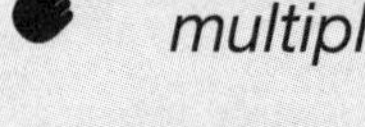

In a mathematical sentence with mixed multiplication and addition, or multiplication and subtraction, do the multiplication first.

Answer these.

3 × 10 + 2 = ☐ 5 × 10 + 5 = ☐ 10 × 9 + 9 = ☐

10 × 7 + 6 = ☐ 6 × 10 + 8 = ☐ 12 × 10 – 9 = ☐

11 × 10 + 6 = ☐ 10 × 10 – 10 = ☐ 8 + 10 × 4 = ☐

Time spent: ______ min ______ sec. Total: ______ out of 40

25 Multiplication of 5

Date:

Day of Week:

STEP 1 (1 min) Warm-up

Start the timer

Answer these.

4 × 10 =	5 + 25 =	10 × 8 =	60 – 5 =
10 × 10 =	7 × 10 =	40 – 5 =	20 + 5 =
5 + 5 + 5 =	5 + 5 =		

STEP 2 (2.5 min) Rapid calculation

Answer these.

1 × 5 =	2 × 5 =	3 × 5 =	4 × 5 =
5 × 5 =	6 × 5 =	7 × 5 =	8 × 5 =
9 × 5 =	5 × 10 =	5 × 1 =	5 × 2 =
5 × 3 =	5 × 4 =	5 × 0 =	5 × 7 =
5 × 11 =	5 × 8 =	5 × 9 =	12 × 5 =

STEP 3 (1.5 min) Challenge

Answer these.

5 × 10 =	4 × 5 =	5 × 2 =	5 × 4 =
11 × 10 =	10 × 10 =	8 × 5 =	5 × 0 =
8 × 10 =	0 × 10 =	0 × 5 =	9 × 5 =

Time spent: ______ min ______ sec. Total: ______ out of 42

Date: ______________________

Day of Week: ______________________

STEP 1 (1 min) Warm-up

Start the timer

Answer these.

$30 + 25 = \square$	$45 - 20 = \square$	$35 + 60 = \square$	$15 + 60 = \square$
$65 - 45 = \square$	$55 - 35 = \square$	$25 + 60 = \square$	$70 - 15 = \square$
$30 + 10 = \square$	$60 - 5 = \square$	$25 + 25 = \square$	$100 - 35 = \square$

STEP 2 (2.5 min) Rapid calculation

Fill in the missing numbers.

$10 \times 4 = \square$	$12 \times 5 = \square$	$3 \times 5 = \square$	$10 \times 8 = \square$
$7 \times 5 = \square$	$9 \times 10 = \square$	$5 \times 6 = \square$	$10 \times 11 = \square$
$8 \times 5 = \square$	$5 \times 3 = \square$	$5 \times 10 = \square$	$5 \times 5 = \square$
$10 \times 9 = \square$	$1 \times 5 = \square$	$9 \times 5 = \square$	$10 \times 0 = \square$
$5 \times \square = 45$	$\square \times 10 = 120$	$5 \times \square = 55$	$\square \times 10 = 40$

STEP 3 (1.5 min) Challenge

Fill in the missing numbers.

$10 \times 11 - 7 = \square$	$10 \times 5 + 33 = \square$	$6 \times 5 + 11 = \square$
$5 \times 12 - 12 = \square$	$8 \times 5 + 24 = \square$	$0 \times 10 + 10 = \square$
$\square - 4 \times 5 = 15$	$45 - 5 \times \square = 35$	$5 \times 12 + \square = 100$

Time spent: ________ min ________ sec. Total: ________ out of 41

27 Multiplication of 2

Date: ______________________

Day of Week: ______________________

STEP 1 (1 min) Warm-up

Start the timer

Answer these.

0 × 2 = ☐ 10 – 2 = ☐ 4 × 2 = ☐ 11 × 2 = ☐

5 + 5 = ☐ 2 × 2 = ☐ 16 – 2 = ☐ 7 + 7 = ☐

2 + 2 + 2 = ☐ 2 + 2 + 2 + 2 = ☐ 22 + 2 = ☐

STEP 2 (2.5 min) Rapid calculation

Start the timer

Answer these.

2 × 4 = ☐ 2 × 5 = ☐ 3 × 2 = ☐ 2 × 8 = ☐

2 × 6 = ☐ 9 × 2 = ☐ 6 × 2 = ☐ 10 × 2 = ☐

8 × 2 = ☐ 2 × 3 = ☐ 2 × 9 = ☐ 2 × 7 = ☐

7 × 2 = ☐ 1 × 2 = ☐ 12 × 2 = ☐ 2 × 0 = ☐

5 × 2 = ☐ 2 × 10 = ☐ 2 × 11 = ☐ 2 × 1 = ☐

STEP 3 (1.5 min) Challenge

Start the timer

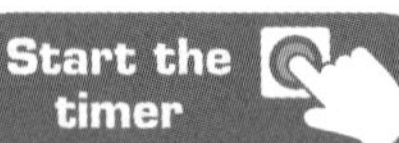

Fill in the missing numbers.

☐ × 7 = 14 10 × ☐ = 30 ☐ × 2 = 16

4 × ☐ = 8 2 × ☐ = 24 ☐ × 5 = 15

☐ × 11 = 22 ☐ × 10 = 110 12 × ☐ = 60

Time spent: ________ min ________ sec. Total: ________ out of 40

Date:

Day of Week:

STEP 1 (1 min) Warm-up

Start the timer 

Answer these.

59 − 38 = □	86 − 52 = □	28 + 24 = □	25 + 16 = □
90 − 29 = □	85 − 27 = □	32 + 18 = □	36 + 45 = □
24 + 24 − 17 = □	63 − 5 × 4 = □		

STEP 2 (2.5 min) Rapid calculation

Start the timer 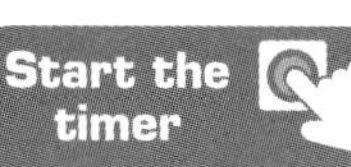

Answer these.

10 × 2 = □	5 × 4 = □	5 × 10 = □	2 × 3 = □
8 × 5 = □	10 × 6 = □	7 × 2 = □	5 × 6 = □
2 × 10 = □	2 × 8 = □	4 × 5 = □	10 × 10 = □
8 × 2 = □	5 × 5 = □	9 × 10 = □	2 × 7 = □
9 × 5 = □	2 × 9 = □	6 × 2 = □	3 × 10 = □

STEP 3 (1.5 min) Challenge

Start the timer 

Fill in the missing numbers.

16 = 2 × □ = 8 × □	15 = □ × 5 = □ × 3	40 = □ × 10 = □ × 4
12 = □ × 6 = □ × 2	35 = □ × 7 = □ × 5	90 = □ × 9 = □ × 10
14 = □ × 7 = □ × 2	45 = 9 × □ = 5 × □	70 = □ × 7 = 7 × □

Time spent: ______ min ______ sec. Total: ______ out of 39

29 Breaking numbers into 2s, 5s and 10s

Date: ____________________

Day of Week: ____________________

STEP 1 (1 min) Warm-up

Start the timer

Answer these.

$8 \times 5 = \square$ $5 \times 2 = \square$ $10 \times 4 = \square$ $2 \times 9 = \square$ $3 \times 5 = \square$

$2 \times 7 = \square$ $6 \times 5 - 16 = \square$ $40 - 2 \times 3 = \square$ $8 \times 5 + 14 = \square$

STEP 2 (2.5 min) Rapid calculation

Fill in the missing numbers.

1. $4 - 2 - 2 = \square$ $4 = \square \times 2$
2. $20 - 5 - 5 - 5 - 5 = \square$ $20 = \square \times 4$
3. $15 - 5 - 5 - 5 = \square$ $15 = \square \times 3$
4. $40 - 10 - 10 - 10 - 10 = \square$ $40 = \square \times 4$
5. $8 - 2 - 2 - 2 - 2 = \square$ $8 = \square \times 4$
6. $25 - 5 - 5 - 5 - 5 - 5 = \square$ $25 = \square \times 5$
7. $30 - 10 - 10 - 10 = \square$ $30 = \square \times 3$
8. $10 - 2 - 2 - 2 - 2 - 2 = \square$ $10 = \square \times 5$
9. $50 - 10 - 10 - 10 - 10 - 10 = \square$ $50 = \square \times 5$

STEP 3 (1.5 min) Challenge

Fill in the missing numbers.

$6 = 2 \times \square$ $10 = 5 \times \square$ $20 = 10 \times \square$ $14 = \square \times 7$

$35 = 5 \times \square$ $60 = \square \times 6$ $2 = \square \times 1$ $5 = 5 \times \square$

$10 = 10 \times \square$ $6 = \square \times 3$ $15 = 5 \times \square$ $40 = 10 \times \square$

Time spent: ________ min ________ sec. Total: ________ out of 39

Date: ____________________

Day of Week: ____________________

Finding quotients using multiplication facts

STEP 1 (1 min) Warm-up

Start the timer 

Answer these.

$2 \times 8 = \square$ $8 \times 5 = \square$ $10 \times 9 = \square$ $6 \times 5 = \square$

$2 \times 2 + 7 = \square$ $5 \times 4 - 6 = \square$ $4 \times 10 + 15 = \square$ $30 - 7 \times 2 = \square$

STEP 2 (2.5 min) Rapid calculation

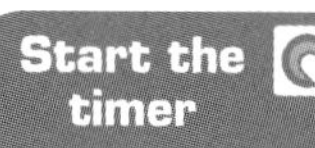

Answer these.

1. $3 \times 2 = \square$ $6 \div 3 = \square$ $6 \div 2 = \square$

2. $2 \times 4 = \square$ $8 \div 2 = \square$ $8 \div 4 = \square$

3. $5 \times 4 = \square$ $20 \div 5 = \square$ $20 \div 4 = \square$

4. $2 \times 9 = \square$ $18 \div 2 = \square$ $18 \div 9 = \square$

5. $8 \times 5 = \square$ $40 \div 8 = \square$ $40 \div 5 = \square$

6. $7 \times 10 = \square$ $70 \div 7 = \square$ $70 \div 10 = \square$

7. $3 \times 5 = \square$ $15 \div 3 = \square$ $15 \div 5 = \square$

8. $6 \times 10 = \square$ $60 \div 10 = \square$ $60 \div 6 = \square$

9. $7 \times 5 = \square$ $35 \div 5 = \square$ $35 \div 7 = \square$

10. $10 \times 8 = \square$ $80 \div 10 = \square$ $80 \div 8 = \square$

11. $9 \times 5 = \square$ $45 \div 5 = \square$ $45 \div 9 = \square$

12. $2 \times 10 = \square$ $20 \div 2 = \square$ $20 \div 10 = \square$

STEP 3 (1.5 min) Challenge

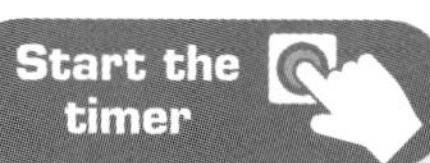

Answer these.

$10 \div 5 = \square$ $12 \div 2 = \square$ $40 \div 10 = \square$ $16 \div 2 = \square$

$70 \div 10 = \square$ $14 \div 2 = \square$ $40 \div 5 = \square$ $100 \div 10 = \square$

Time spent: ______ min ______ sec. Total: ______ out of 52

31 Division with dividend 0

Date: ____________

Day of Week: ____________

STEP 1 (1 min) Warm-up

Start the timer

Answer these.

$4 \times 5 =$ ☐ $3 \times 10 =$ ☐ $8 \times 2 =$ ☐

$10 \times 10 =$ ☐ $10 \div 5 =$ ☐ $2 \times 2 =$ ☐

$10 \times 0 =$ ☐ $40 \div 5 =$ ☐

STEP 2 (2.5 min) Rapid calculation

Answer these.

$0 \div 5 =$ ☐ $0 \times 2 =$ ☐ $0 \div 10 =$ ☐ $2 \times 4 =$ ☐

$5 \div 5 =$ ☐ $0 \div 2 =$ ☐ $2 \div 2 =$ ☐ $10 \div 10 =$ ☐

$0 \times 10 =$ ☐ $5 \times 0 =$ ☐ $23 + 35 \div 5 =$ ☐ $0 \times 10 + 0 =$ ☐

$94 - 6 \times 10 =$ ☐ $0 \times 2 + 77 =$ ☐ $1 \div 1 =$ ☐ $0 \div 5 + 15 =$ ☐

STEP 3 (1.5 min) Challenge

Fill in each box with **>**, **<** or **=**.

$36 + 0$ ☐ $36 - 0$ $10 - 0$ ☐ $0 \div 10$ 0×5 ☐ $0 \div 5$

10×0 ☐ $10 + 0$ $10 + 10$ ☐ 10×10 $5 - 5$ ☐ $5 \div 5$

$0 + 2$ ☐ $0 \div 2$ $2 - 2$ ☐ $0 \div 2$ 0×10 ☐ $10 \div 10$

Time spent: ______ min ______ sec. Total: ______ out of 33

Date: ____________________

Day of Week: ____________________

Practice 32

STEP 1 (1 min) Warm-up

Start the timer

Answer these.

$4 \times 5 =$ □	$25 + 4 =$ □	$10 \times 5 =$ □	$0 \times 2 =$ □
$39 + 21 =$ □	$46 - 32 =$ □	$26 + 14 =$ □	$6 \times 10 =$ □

STEP 2 (2.5 min) Rapid calculation

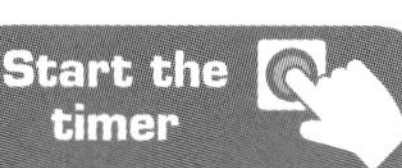

Answer these.

$5 \times 7 =$ □	$0 \div 2 =$ □	$35 \div 5 =$ □	$4 \times 10 =$ □
$50 \div 5 =$ □	$0 \div 10 =$ □	$0 \times 10 =$ □	$5 \times 7 =$ □
$70 \div 10 =$ □	$2 \times 7 =$ □	$1 \times 5 =$ □	$10 \times 0 =$ □
$100 \div 10 =$ □	$20 \div 2 =$ □	$30 \div 5 =$ □	$2 \times 10 =$ □
$40 \div 5 =$ □	$8 \times 2 =$ □	$90 \div 10 =$ □	$0 \div 5 =$ □

STEP 3 (1.5 min) Challenge

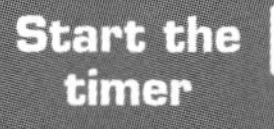

Fill in the missing numbers.

$50 \div$ □ $= 10$	□ $\div 7 = 5$	$15 \div$ □ $= 3$
□ $\div 10 = 9$	□ $- 8 = 6$	□ $\times 5 = 40$
□ $\times 2 = 0$	□ $\div 2 = 10$	□ $+ 10 = 26$

Time spent: ______ min ______ sec. Total: ______ out of 37

33 Commutative law of multiplication

Date: ______________________

Day of Week: ______________________

STEP 1 (1 min) Warm-up

Start the timer

Answer these.

1. $2 \times 6 =$ ☐

$6 \times 2 =$ ☐

2. $4 \times 5 =$ ☐

$5 \times 4 =$ ☐

3. $7 \times 10 =$ ☐

$10 \times 7 =$ ☐

4. $8 \times 2 =$ ☐

$2 \times 8 =$ ☐

5. $9 \times 5 =$ ☐

$5 \times 9 =$ ☐

6. $10 \times 9 =$ ☐

$9 \times 10 =$ ☐

STEP 2 (2.5 min) Rapid calculation

Fill in the missing numbers.

$5 \times 3 = 3 \times$ ☐ $=$ ☐

$10 \times 4 = 4 \times$ ☐ $=$ ☐

$2 \times 7 = 7 \times$ ☐ $=$ ☐

$10 \times 6 = 6 \times$ ☐ $=$ ☐

$5 \times 7 = 7 \times$ ☐ $=$ ☐

$2 \times 9 = 9 \times$ ☐ $=$ ☐

$10 \times 8 =$ ☐ $\times$ ☐ $=$ ☐

$4 \times 2 =$ ☐ $\times$ ☐ $=$ ☐

$5 \times 8 =$ ☐ $\times$ ☐ $=$ ☐

$10 \times 2 =$ ☐ $\times$ ☐ $=$ ☐

$5 \times 6 =$ ☐ $\times$ ☐ $=$ ☐

$10 \times 5 =$ ☐ $\times$ ☐ $=$ ☐

STEP 3 (1.5 min) Challenge

Fill in the missing numbers.

$9 \times$ ☐ $=$ ☐ $=$ ☐ $\times 5$

$8 \times$ ☐ $=$ ☐ $=$ ☐ $\times 2$

$3 \times$ ☐ $=$ ☐ $=$ ☐ $\times 10$

$6 \times$ ☐ $=$ ☐ $=$ ☐ $\times 2$

$9 \times$ ☐ $=$ ☐ $=$ ☐ $\times 10$

$7 \times$ ☐ $=$ ☐ $=$ ☐ $\times 5$

$7 \times$ ☐ $=$ ☐ $=$ ☐ $\times 10$

$8 \times$ ☐ $=$ ☐ $=$ ☐ $\times 5$

Time spent: ________ min ________ sec. Total: ________ out of 32

Date: ____________________

Day of Week: ____________________

Subtraction and division are not commutative

STEP 1 (1 min) Warm-up

Start the timer

Do these calculations have the same answer? Write **yes** or **no** next to each one.

$12 + 5$ and $5 + 12$ ____________

5×7 and 7×5 ____________

$10 - 3$ and $3 - 10$ ____________

$10 \div 2$ and $2 \div 10$ ____________

$18 - 6$ and $6 - 18$ ____________

8×2 and 2×8 ____________

$40 \div 5$ and $5 \div 40$ ____________

$33 + 50$ and $50 + 33$ ____________

STEP 2 (2.5 min) Rapid calculation

1. Fill in the missing numbers.

$13 + 6 = 6 + \square = \square$

$7 \times 5 = 5 \times \square = \square$

$26 + 30 = 30 + \square = \square$

$9 \times 10 = 10 \times \square = \square$

$34 + 8 = 8 + \square = \square$

$2 \times 7 = 7 \times \square = \square$

2. Fill in the missing numbers. The first one has been done for you.

$12 - 9$ does not equal $9 -$ **12**

$50 \div \square$ does not equal $5 \div 50$

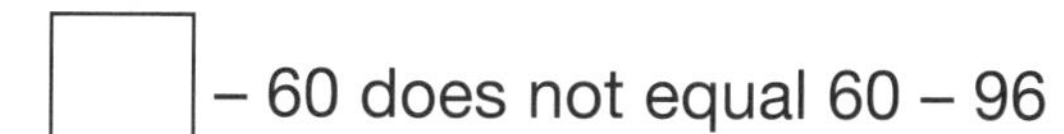

$\square - 60$ does not equal $60 - 96$

$15 \div 5$ does not equal $5 \div \square$

$20 \div \square$ does not equal $2 \div 20$

$\square \div 10$ does not equal $10 \div 60$

STEP 3 (1.5 min) Challenge

Fill in the missing numbers so that the four calculations in each set give the same answer.

1. $\square + 6 = 23 - \square = 2 \times \square = 20 \div \square = 10$

2. $\square + 0 = 83 - \square = 4 \times \square = 80 \div \square = 8$

3. $4 + \square = \square - 58 = 1 \times \square = \square \div 5 = 4$

4. $\square + 1 = \square - 77 = 3 \times \square = \square \div 2 = 6$

Time spent: ________ min ________ sec. Total: ________ out of 23

35 General review (5)

Date: ______________

Day of Week: ______________

STEP 1 (1 min) Warm-up

Start the timer

Answer these.

1. 14 + 9 = ☐

23 – 9 = ☐

23 – 14 = ☐

2. 27 + 6 = ☐

33 – 6 = ☐

33 – 27 = ☐

3. 39 + 5 = ☐

44 – 5 = ☐

44 – 39 = ☐

4. 27 + 36 = ☐

63 – 36 = ☐

63 – 27 = ☐

STEP 2 (2.5 min) Rapid calculation

Start the timer

Answer these.

51 – 37 = ☐	33 – 26 = ☐	45 + 28 = ☐	65 + 27 = ☐
34 + 48 = ☐	96 – 48 = ☐	27 + 39 = ☐	84 – 36 = ☐
72 – 58 = ☐	18 + 49 = ☐	69 + 23 = ☐	79 + 14 = ☐
63 – 49 = ☐	93 – 47 = ☐	36 + 49 = ☐	58 + 25 = ☐
52 – 24 = ☐	18 + 38 = ☐	83 – 54 = ☐	55 – 26 = ☐

STEP 3 (1.5 min) Challenge

Start the timer

Fill in the missing numbers.

☐ + 67 = 81	21 + ☐ = 40	92 = 94 – ☐
☐ – 24 = 75	85 – ☐ = 38	☐ + 38 = 65
100 – ☐ = 31	☐ – 35 = 48	78 + ☐ + 12 = 100

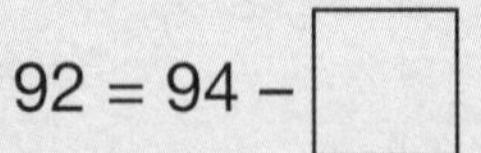

Time spent: ________ min ________ sec. Total: ________ out of 41

Date: ______________________

Day of Week: ______________________

General review (6) 36

STEP 1 (1 min) Warm-up

Answer these.

37 + 24 = 40 + ☐ 63 + 37 = ☐ + 40 44 + 38 = ☐ + 40

29 + 42 = 30 + ☐ 47 + 46 = 50 + ☐

66 − 27 = ☐ − 30 41 − 22 = ☐ − 20 74 − 58 = ☐ − 60

45 − 29 = ☐ − 30 91 − 23 = ☐ − 20

STEP 2 (2.5 min) Rapid calculation

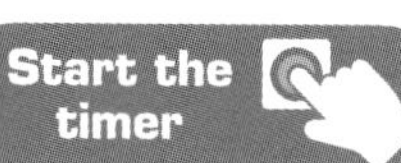

Fill in the missing numbers. Some have been done for you.

46 + 46 = **50** + **42** = ☐ 37 + 48 = ☐ + ☐ = ☐

24 + 67 = ☐ + ☐ = ☐ 52 + 19 = ☐ + ☐ = ☐

29 + 63 = ☐ + ☐ = ☐ 76 + 14 = ☐ + ☐ = ☐

74 − 46 = **78** − **50** = ☐ 53 − 24 = ☐ − ☐ = ☐

75 − 57 = ☐ − ☐ = ☐ 72 − 23 = ☐ − ☐ = ☐

64 − 36 = ☐ − ☐ = ☐ 82 − 48 = ☐ − ☐ = ☐

STEP 3 (1.5 min) Challenge

Answer these.

86 − 48 = ☐ 44 − 17 = ☐ 76 − 37 = ☐ 73 − 59 + 28 = ☐

68 + 14 = ☐ 35 + 36 = ☐ 44 + 29 = ☐ 97 − 43 − 25 = ☐

64 + 28 − 53 = ☐ 15 + 24 + 36 = ☐

Time spent: ________ min ________ sec. Total: ________ out of 32

37 General review (7)

Date:

Day of Week:

STEP 1 (1 min) Warm-up

Start the timer

Answer these.

$32 + 16 =$ $32 - 16 =$ $7 \times 5 =$ $8 \times 8 =$

$15 \div 5 =$ $20 \div 2 =$ $100 - 26 =$ $81 - 39 =$

$24 + 37 =$ $80 \div 10 =$ $30 \div 5 =$ $10 \times 7 =$

STEP 2 (2.5 min) Rapid calculation

Answer these.

$10 \times 4 =$ $8 \times 5 =$ $35 \div 5 =$ $14 \div 2 =$

$4 \times 5 =$ $8 \times 2 =$ $25 \div 5 =$ $90 \div 10 =$

$8 \times 10 =$ $2 \times 5 =$ $45 \div 5 =$ $30 \div 10 =$

$8 \times 5 =$ $6 \times 10 =$ $18 \div 2 =$ $40 \div 5 =$

$5 \times 9 =$ $10 \times 10 =$ $12 \div 2 =$ $70 \div 10 =$

STEP 3 (1.5 min) Challenge

Answer these.

$2 \times 10 + 31 =$ $30 \div 5 - 2 =$ $2 \times 8 + 15 =$

$5 \times 10 + 28 =$ $10 \div 2 + 9 =$ $9 \times 5 - 19 =$

$10 \times 4 + 59 =$ $40 \div 5 - 7 =$ $0 \div 6 + 78 =$

Time spent: ______ min ______ sec. Total: ______ out of 41

Date:

Day of Week:

STEP 1 (1 min) Warm-up

Start the timer

Answer these.

$27 + 16 =$	$39 - 15 =$	$4 \times 2 =$	$7 \times 2 =$
$50 \div 5 =$	$40 \div 10 =$	$34 - 17 =$	$31 - 12 =$
$26 + 49 =$	$12 \div 2 =$	$50 \div 10 =$	$5 \times 5 =$

STEP 2 (2.5 min) Rapid calculation

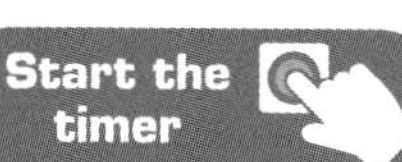

Answer these.

$10 \times 3 =$	$40 \div 5 =$	$50 \div 10 =$	$20 \div 5 =$
$10 \times 6 =$	$10 \div 5 =$	$6 \div 2 =$	$60 \div 10 =$
$1 \times 5 =$	$100 \div 10 =$	$2 \div 2 =$	$5 \div 5 =$
$82 - 19 =$	$36 + 47 =$	$35 + 65 =$	$0 \div 10 =$
$3 + 59 =$	$100 - 63 =$	$41 - 36 =$	$53 - 28 =$

STEP 3 (1.5 min) Challenge

Answer these.

$7 \times 5 + 35 =$	$10 \times 10 - 45 =$	$10 \times 2 - 18 =$
$8 \times 5 - 22 =$	$4 \times 10 + 48 =$	$5 \times 9 + 45 =$
$5 \times 0 + 91 =$	$0 \div 10 + 18 =$	$2 \times 5 \times 10 =$

Time spent: ______ min ______ sec. Total: ______ out of 41

39 General review (9)

Date: ____________________

Day of Week: ____________________

STEP 1 (1 min) Warm-up

Start the timer

Potted plants are arranged in equal rows. How many plants are there in total?

	Total number of plants
4 rows of 5 plants	
3 rows of 10 plants	
5 rows of 8 plants	

	Total number of plants
6 rows of 2 plants	
2 rows of 9 plants	
10 rows of 9 plants	

STEP 2 (2.5 min) Rapid calculation

Start the timer

Which has more beads? Circle the correct answer in each pair.

2 bags of 8 beads **or** 3 bags of 5 beads

9 bags of 2 beads **or** 2 bags of 10 beads

6 bags of 2 beads **or** 1 bag of 10 beads

5 bags of 10 beads **or** 10 bags of 4 beads

5 bags of 8 beads **or** 3 bags of 10 beads

4 bags of 10 beads **or** 7 bags of 5 beads

5 bags of 9 beads **or** 8 bags of 5 beads

5 bags of 7 beads **or** 8 bags of 5 beads

7 bags of 2 beads **or** 2 bags of 8 beads

5 bags of 5 beads **or** 10 bags of 3 beads

STEP 3 (1.5 min) Challenge

1. There are 10 groups of 5 children. How many more children will make 63? ☐

2. There are 8 groups of 2 children. How many more children will make 31? ☐

3. There are 7 groups of 10 children. How many more children will make 94? ☐

4. There are 5 groups of 9 children. How many more children will make 68? ☐

Time spent: ________ min ________ sec. Total: ________ out of 20

Date: ____________

Day of Week: ____________

STEP 1 (1 min) Warm-up

Start the timer

How many counters are in each tray?

20 counters shared equally between 2 trays ☐

18 counters shared equally between 9 trays ☐

60 counters shared equally between 10 trays ☐

90 counters shared equally between 10 trays ☐

30 counters shared equally between 5 trays ☐

45 counters shared equally between 5 trays ☐

STEP 2 (2.5 min) Rapid calculation

Start the timer

Seeds are planted in equal rows. How many seeds are in each row?

35 seeds in 5 rows ☐

80 seeds in 10 rows ☐

16 seeds in 2 rows ☐

50 seeds in 5 rows ☐

12 seeds in 6 rows ☐

15 seeds in 3 rows ☐

40 seeds in 4 rows ☐

20 seeds in 2 rows ☐

20 seeds in 4 rows ☐

100 seeds in 10 rows ☐

STEP 3 (1.5 min) Challenge

Start the timer

1. £50 is shared equally between 5 boys. Each boy needs £ ☐ more to have £13.
2. £60 is shared equally between 10 girls. Each girl needs £ ☐ more to have £24.
3. £18 is shared equally between 2 boys. Each boy needs £ ☐ more to have £31.
4. £45 is shared equally between 9 girls. Each girl needs £ ☐ more to have £42.
5. £90 is shared equally between 9 boys. Each boy needs £ ☐ more to have £73.
6. £16 is shared equally between 8 girls. Each girl needs £ ☐ more to have £62.

Time spent: ______ min ______ sec. Total: ______ out of 22

ANSWERS

Answers are given from top left, left to right, unless otherwise stated.

Test 1

Step 1:

13; 9; 15; 2; 6; 13; 20; 10; 17; 5

Step 2:

20; 9; 17; 18; 12; 21; 20; 8; 18; 12; 11; 0; 15; 19; 7; 14; 8; 15; 5; 10

Step 3:

8; 11; 9; 8; 2; 6; 14; 6

Test 2

Step 1:

15; 10; 19; 10; 14; 15; 13; 10; 7; 7

Step 2:

<; >; =; <; <; <; =; >; <; =; >; >; =; =; >; <

Step 3:

1; 9; 9; 4; 13; 5; 15; 10

Test 3

Step 1:

70; 90; 80; 90; 72; 95; 84; 97

Step 2:

72; 95; 84; 97; 97; 88; 93; 77; 85; 91; 49; 82; 72; 55; 49; 55; 83; 86; 71; 100

Step 3:

30; 40; 20; 48; 60; 29; 30; 35; 48

Test 4

Step 1:

First column: £17; £46; £83; £59; £65

Second column: 36p; 48p; 92p; 97p; 70p

Step 2:

1. First column: 57 m; 75 m; 85 m; 63 m
   Second column: 67 cm; 84 cm; 82 cm; 65 cm
2. <; =; <; >; =; >; =; <

Step 3:

20 m; 42 m; 40 m; 30 cm; 49 cm;
40 cm; 10 m; 20 cm; 56 m

Test 5

Step 1:

40; 70; 20; 30; 41; 77; 26; 34

Step 2:

41; 77; 26; 34; 57; 28; 33; 27; 45; 11; 29; 22; 2; 45; 45; 5; 7; 11; 31; 0

Step 3:

33; 78; 82; 70; 40; 96; 50; 30; 57

Test 6

Step 1:

First column: £25; £36; £58; £64; £41

Second column: 57p; 43p; 28p; 15p; 14p

Step 2:

1. First column: 12 m; 38 m; 59 m; 11 m
   Second column: 27 cm; 26 cm; 5 cm; 19 cm
2. <; >; =; <; <; >; =; <

Step 3:

26 m; 27 m; 69 m; 53 cm; 62 cm;
93 cm; 43 m; 18 cm; 55 m

Test 7

Step 1:

8; 9; 7; 9; 60; 80; 80; 47

Step 2:

89; 55; 79; 69; 58; 49; 89; 39; 38; 69; 57; 46; 68; 29; 47; 59; 68; 29; 47; 59

Step 3:

60; 90; 60; 71; 51; 43; 40; 81; 6

Test 8

Step 1:

3; 1; 5; 2; 53; 71; 45; 92

Step 2:

14; 23; 39; 45; 51; 77; 58; 69; 59; 59; 59; 59; 52; 54; 56; 58; 53; 71; 45; 92

Step 3:

42; 77; 38; 55; 56; 90; 88; 68; 2

Test 9

Step 1:

61; 37; 51; 84; 92; 28; 49; 66

Step 2:

59; 68; 35; 68; 34; 34; 73; 81; 70; 70; 70; 60; 66; 48; 30; 70; 77; 35

Step 3:

5; 40; 5; 73; 9; 7; 6; 7; 82

Test 10

Step 1:

7; 8; 80; 80; 90; 90; 40; 86

Step 2:

91; 87; 85; 84; 95; 85; 45; 79; 67; 96; 97; 84; 42; 86; 68; 26; 89; 97; 98; 46

Step 3:

33; 55; 77; 55; 77; 99; 70; 28; 22

Test 11

Step 1:
10; 60; 50; 20; 20; 12; 64; 53

Step 2:
46; 20; 40; 80; 13; 61; 33; 42; 71; 45; 65; 23; 53; 0; 39; 41; 25; 21

Step 3:
66; 44; 22; 73; 53; 33; 50; 88; 41

Test 12

Step 1:
73; 19; 26; 97; 95; 21; 87; 94

Step 2:
99; 30; 88; 50; 90; 90; 21; 27; 81; 55; 90; 65; 0; 67; 18; 78; 72; 80

Step 3:
49; 51; 29; 63; 26; 36; 15; 52; 70

Test 13

Step 1:
First column: £50; £40; £70; £70; £90
Second column: 19p; 27p; 29p; 39p; 5p

Step 2:
First column: 58 m; 38 m; 48 cm; 15 cm; 90 m; 84 m; 91 cm; 57 cm; 65 cm
Second column: 20 m; 93 m; 75 cm; 75 cm; 19 m; 70 m; 78 cm; 21 cm; 55 cm

Step 3:
£60; £45; £33; £72; £18; £38; 61p; 87p; 24p

Test 14

Step 1:
13; 13; 11; 12; 13; 16; 15; 14

Step 2:
11; 12; 12; 11; 19; 18; 14; 15; 17; 18; 19; 17; 15; 18; 15; 16; 13; 12

Step 3:
25; 25; 24; 20; 22; 27; 21; 24

Test 15

Step 1:
35; 88; 79; 87; 5; 62; 34; 40

Step 2:
72; 96; 27; 19; 44; 78; 62; 71; 26; 14; 17; 28; 4; 81; 23; 66; 91; 5

Step 3:
22; 22; 26; 18; 28; 53; 16; 63; 75

Test 16

Step 1:
62; 26; 85; 8; 34; 33; 65; 39

Step 2:
31; 49; 28; 39; 54; 15; 27; 47; 38; 25; 29; 48; 66; 59; 34; 36

Step 3:
<; >; =; >; <; =; >; <

Test 17

Step 1:
£21; £18; £15; £18; 13p; 20p; 18p; 17p

Step 2:
>; <; >; <; = >; =; <

Step 3:
£47; £68; £95; £29; 19p; 71p; 52p; 83p

Test 18

Step 1:
71 km; 72 km; 74 km; 72 km; 93 km; 94 km

Step 2:
£20; £18; £20; £20; £26; £28; £24; £19; £17; £22

Step 3:
1. £37
2. £48
3. £38
4. £67

Test 19

Step 1:
85, 85; 95, 95; 92, 92; 92, 92; 94, 94; 86, 86

Step 2:
1. 34 + 25 = 59
2. 16 + 67 = 83
3. 37 + 58 = 95
4. 25 + 47 = 72
5. 24 + 59 = 83
6. 47 + 29 = 76
7. 68 + 23 = 91
8. 18 + 75 = 93
9. 35 + 43 = 78
10. 68 + 18 = 86
11. 16 + 83 = 99
12. 44 + 37 = 81

Step 3:
1. 83; 83; 83; 83
2. 86; 86; 86; 27, 86
3. 98; 98; 98; 58, 98
4. 91; 42, 91; answers will vary; answers will vary

Answers

Test 20

Step 1:

1. Circled: 13, 21, 35, 47, 49, 85, 79, 57
2. Circled: 88, 46, 54, 42, 92, 98, 34, 64, 74, 70, 2, 100

Step 2:

1. 5, 7, 9; 15, 17, 19; 33, 35, 37; 51, 53, 55, 57, 59; 77, 79, 81, 83; 85, 87, 89, 91
2. 2, 4, 6; 14, 16, 18; 26, 28; 42, 44, 46, 48; 66, 68, 70, 72; 90, 92, 94, 96

Step 3:

1. 82, 27 **2.** 68, 17 **3.** 64, 15 **4.** 80, 43

Test 21

Step 1:

6; 9; 16; 20; 18; 21; 12; 24; 12; 27; 28; 40

Step 2:

4 × 2 = 8	3 × 4 = 12	4 × 3 = 12	5 × 1 = 5
4 × 5 = 20	3 × 7 = 21	3 × 9 = 27	4 × 8 = 32
5 × 6 = 30	5 × 7 = 35	7 × 1 = 7	3 × 3 = 9
8 × 5 = 40	2 × 8 = 16		

Step 3:

9 × 3 = 27; 7 × 8 = 56; 4 × 10 = 40; 8 × 9 = 72; 8 × 12 = 96

Test 22

Step 1:

6; 20; 40; 15; 30; 8; 50; 12; 30; 70

Step 2:

five 2s, 5 times 2, 5 × 2 = 10;
four 10s, 4 times 10, 4 × 10 = 40;
eight 2s, 8 times 2, 8 × 2 = 16;
six 10s, 6 times 10, 6 × 10 = 60;
five 5s, 5 times 5, 5 × 5 = 25

Step 3:

10, 2 × 5 = 10, 5 × 2 = 10;
30, 3 × 10 = 30, 10 × 3 = 30;
10, 5 × 2 = 10, 2 × 5 = 10;
35, 7 × 5 = 35, 5 × 7 = 35;
70, 7 × 10 = 70, 10 × 7 = 70;
40, 8 × 5 = 40, 5 × 8 = 40;
20, 10 × 2 = 20, 2 × 10 = 20

Test 23

Step 1:

3, 5, 15
2, 10, 20
2, 5, 10
5, 4, 20
10, 5, 50

Step 2:

2 × 5 = 10, 5 × 2 = 10;
3 × 10 = 30, 10 × 3 = 30;
2 × 7 = 14, 7 × 2 = 14;
4 × 5 = 20, 5 × 4 = 20;
5 × 10 = 50, 10 × 5 = 50

Step 3:

8; 30; 70; 18; 40; 90

Test 24

Step 1:

20; 70; 50; 80; 50; 60; 70; 80; 30; 20; 40

Step 2:

10; 20; 30; 40; 50; 60; 70; 80; 90; 100; 10; 20; 30; 40; 50; 60; 110; 80; 120; 0

Step 3:

32; 55; 99; 76; 68; 111; 116; 90; 48

Test 25

Step 1:

40; 30; 80; 55; 100; 70; 35; 25; 15; 10

Step 2:

5; 10; 15; 20; 25; 30; 35; 40; 45; 50; 5; 10; 15; 20; 0; 35; 55; 40; 45; 60

Step 3:

50; 20; 10; 20; 110; 100; 40; 0; 80; 0; 0; 45

Test 26

Step 1:

55; 25; 95; 75; 20; 20; 85; 55; 40; 55; 50; 65

Step 2:

40; 60; 15; 80; 35; 90; 30; 110; 40; 15; 50; 25; 90; 5; 45; 0; 9; 12; 11; 4

Step 3:

103; 83; 41; 48; 64; 10; 35; 2; 40

Test 27

Step 1:

0; 8; 8; 22; 10; 4; 14; 14; 6; 8; 24

Step 2:

8; 10; 6; 16; 12; 18; 12; 20; 16; 6; 18; 14; 14; 2; 24; 0; 10; 20; 22; 2

Step 3:

2; 3; 8; 2; 12; 3; 2; 11; 5

Test 28

Step 1:

21; 34; 52; 41; 61; 58; 50; 81; 31; 43

Step 2:

20; 20; 50; 6; 40; 60; 14; 30; 20; 16; 20; 100; 16; 25; 90; 14; 45; 18; 12; 30

Step 3:

8, 2; 3, 5; 4, 10; 2, 6; 5, 7; 10, 9; 2, 7; 5, 9; 10, 10

Test 29

Step 1:

40; 10; 40; 18; 15; 14; 14; 34; 54

Step 2:

1. 0, 2
2. 0, 5
3. 0, 5
4. 0, 10
5. 0, 2
6. 0, 5
7. 0, 10
8. 0, 2
9. 0, 10

Step 3:

3; 2; 2; 2; 7; 10; 2; 1; 1; 2; 3; 4

Test 30

Step 1:

16; 40; 90; 30; 11; 14; 55; 16

Step 2:

1. 6, 2, 3
2. 8, 4, 2
3. 20, 4, 5
4. 18, 9, 2
5. 40, 5, 8
6. 70, 10, 7
7. 15, 5, 3
8. 60, 6, 10
9. 35, 7, 5
10. 80, 8, 10
11. 45, 9, 5
12. 20, 10, 2

Step 3:

2; 6; 4; 8; 7; 7; 8; 10

Test 31

Step 1:

20; 30; 16; 100; 2; 4; 0; 8

Step 2:

0; 0; 0; 8; 1; 0; 1; 1; 0; 0; 30; 0; 34; 77; 1; 15

Step 3:

=; >; =; <; <; <; >; =; <

Test 32

Step 1:

20; 29; 50; 0; 60; 14; 40; 60

Step 2:

35; 0; 7; 40; 10; 0; 0; 35; 7; 14; 5; 0; 10; 10; 6; 20; 8; 16; 9; 0

Step 3:

5; 35; 5; 90; 14; 8; 0; 20; 16

Test 33

Step 1:

1. 12, 12
2. 20, 20
3. 70, 70
4. 16, 16
5. 45, 45
6. 90, 90

Step 2:

5, 15; 10, 40; 2, 14; 10, 60; 5, 35; 2, 18; 8, 10, 80; 2, 4, 8; 8, 5, 40; 2, 10, 20; 6, 5, 30; 5, 10, 50

Step 3:

5, 45, 9; 2, 16, 8; 10, 30, 3; 2, 12, 6; 10, 90, 9; 5, 35, 7; 10, 70, 7; 5, 40, 8

Test 34

Step 1:

yes; yes; no; no; no; yes; no; yes

Step 2:

1. 13, 19; 7, 35; 26, 56; 9, 90; 34, 42; 2, 14
2. 5; 96; 15; 2; 60

Step 3:

1. 4, 13, 5, 2
2. 8, 75, 2, 10
3. 0, 62, 4, 20
4. 5, 83, 2, 12

Test 35

Step 1:

1. 23, 14, 9
2. 33, 27, 6
3. 44, 39, 5
4. 63, 27, 36

Step 2:

14; 7; 73; 92; 82; 48; 66; 48; 14; 67; 92; 93; 14; 46; 85; 83; 28; 56; 29; 29

Step 3:

14; 19; 2; 99; 47; 27; 69; 83; 10

Test 36

Step 1:

21; 60; 42; 41; 43; 69; 39; 76; 46; 88

Step 2:

92; 35 + 50 = 85; 21 + 70 = 91; 51 + 20 = 71; 30 + 62 = 92; 80 + 10 = 90; 28; 49 – 20 = 29; 78 – 60 = 18; 69 – 20 = 49; 68 – 40 = 28; 84 – 50 = 34

Other suitable additions/subtractions are possible.

Step 3:

38; 27; 39; 42; 82; 71; 73; 29; 39; 75

Test 37

Step 1:

48; 16; 35; 64; 3; 10; 74; 42; 61; 8; 6; 70

Step 2:

40; 40; 7; 7; 20; 16; 5; 9; 80; 10; 9; 3; 40; 60; 9; 8; 45; 100; 6; 7

Step 3:

51; 4; 31; 78; 14; 26; 99; 1; 78

Answers

Test 38

Step 1:
43; 24; 8; 14; 10; 4; 17; 19; 75; 6; 5; 25

Step 2:
30; 8; 5; 4; 60; 2; 3; 6; 5; 10; 1; 1; 63; 83; 100; 0; 62; 37; 5; 25

Step 3:
70; 55; 2; 18; 88; 90; 91; 18; 100

Test 39

Step 1:
First table, totals from top to bottom: 20; 30; 40
Second table, totals from top to bottom: 12; 18; 90

Step 2:
2 bags of 8 beads; 4 bags of 10 beads; 2 bags of 10 beads; 5 bags of 9 beads; 6 bags of 2 beads; 8 bags of 5 beads; 5 bags of 10 beads; 2 bags of 8 beads; 5 bags of 8 beads; 10 bags of 3 beads

Step 3:
1. 13 **2.** 15
3. 24 **4.** 23

Test 40

Step 1:
10; 2; 6; 9; 6; 9

Step 2:
7; 8; 8; 10; 2; 5; 10; 10; 5; 10

Step 3:
1. £3 **2.** £18
3. £22 **4.** £37
5. £63 **6.** £60